Introduction To Basic Networking

SUCRE RAMÍREZ

ISBN: 1466288094
ISBN-13: 978-1466288096

PROLOGUE

This book will let you have a general concept on data networks and telecommunications basics. You can solve problems with the protocol more used in the world's IP Version 4 Just as well have a point of view on business cases, and will implement based structure cabling networks, it can also help for the industry certifications. Data networks is introduced from a fundamental point of view where chapters go on the basics and to a considerable level that allows the reader to venture into the area with a good foundation and a practical point of view that give confidence to work in the market.

CONTENT

Acknolagement   i

1 Overview of data telecommunications.   1
   Classification, types of networks. Networks:
   Local area networks, wide area network,
   campus network, Internet, Intranet, and
   Extranet. Architectures: OSI, SNA, TCP / IP,
   conection oriented or conetionless, industry
   and owner protocols. Topologies: Star, bus,
   ring and mesh. Design of Local Area
   Networks and Economic Proposals.

2 OSI, Open standar Interconeccion. Reason   25
   for use. Layer Model and Examples.

3 Level 1. Physical. Electricity, attenuation, 31
   synchronous and asynchronous transmission,
   analog and digital signal, modem, multiplexer,
   nyquist theorem, structure cabling and
   examples.

4 Level 2 - LLC and MAC. Witch. Share   48
   Network medium. Operation mode, filtering,
   segmentation and convergence.

5 Level 3 - Network, Routers. Functions, type, 60
   class hierarchy, mask, subnets, and security.

Methods for calculating IP.

6 Level 4 - transport. Data Flow, windowing,   76
   buffering and congestion. Cases.

7 Interoperability. Levels 5, 6 and 7 Model TCP   82
   / IP and OSI compare network. Application
   protocols.

ACKNOLAGEMENT

Thank you my God because you allowed my growth, you are always with me, with your Holy Spirit and Jesus Christ   can do everything, with you I become stronger, Jose Maria who was the inspiration behind this project, Luis Peralta for his contribution, Maria Isabel for the cover, to the memory of My Mother Eudenia Roca, my dad Sucre Héctor Ramírez and my dear wife Ramona Bernard.

1. GENERAL DESCRIPTION OF DATA TELECOMMUNICATIONS

Data and voice telecommunications have been useful in recent times. Measure network of voice analysis is in Earlans (Unit that measures the volume of voice traffic) but for the study of the data there is no scientific unit that measures the volume of data traffic, we can find solutions subjective about that largely, depend on the experience of those who pose. Here we discuss some aspects in that sense, explane the sampling theorem in order to define the bandwidth. Using examples from real life we create comparisons to facilitate understanding. Doing the suggested exercises you be able to analyze internet address (IP) and in less than a minute will get a satisfactory answer on one of the most used protocols in the world. You can also learn how to design and network systems for market. Also Analise tools that allow to study the analysis and solution of problems in data networks.

Clasification of data telecommunications:

A. Network
B. Arquitecture = Protocol = Logical Topology
C. Fisical Topology

A) Networks. For make communication posible is necesary the existence of a transmitter, a receiver and medium. For example: In typical conversation transmitter would be the larynx and the ear receiver and medium the air that separates them.

Figure 1. Typical Conversation

If people with disabilities, for example, deaf or mute: Body is transmitter and receiver the eye. In both cases there may be capacity to transmit and receive the message, not necessarily understanding it means, to ensure effective communication either must have a common language to optain acknolegement

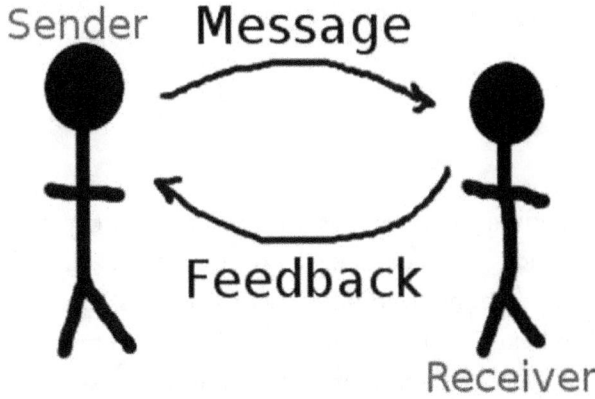

Figure 2. Comunication Element

In the case of equipment that is our interest would be the same; networks are also formed by these three elements. And to ensure effective communication is necessary here protocol standards that are similar to the language that humans use to communicate. The networks are media through which large amounts of data are exchanged. The most common reasons for deciding to install a network are:

• Share programs, files and printer.

• Ability to use network software.

- Creation of working groups.

- Centralized management.

- Access to other operating systems.

- Share resources.

Making a brief analysis we find in the story that the advance of centralized systems and distributed computations need networks for process the data. In 1960 exist Main Frame, centralized systems. In 1970 the Mini computers, Centralized systems and satellites. At 1980 penetrate PCs., Personal Computer. In 1990 distributed systems. Highly dependent on data networks.

MetaFrame combines Distributed System with Centralized System. Transform the way that an organization uses, manage and access applications providing better capacity

management, access and security.
Being able to access to most used
platforms in the market.

Types of Networks:

- LAN, low latency.

- WAN, high latency.

- MAN.

- CAMPUS NETWORK.

- Private, Public, Internet, Intranet and Extranet.

LAN, Local Area Network. This is a system of
communication between computers where the
distance between the equipment should be relatively
small (between 100 and 400 meters, acording to

structu cabling up to 100 meters if we connect three hubs).

WAN, Wide Area Network. Networks with multiple computers on an extended. The latency in these systems is making it difficult to design due to the delays that these causes can vary from a few milliseconds to tens of seconds. It is a term denoting an expanded definition delay, attenuation, caused by slow computers, a lot of software, the level difference, etc.

Man. Metropolitan Area Network. This network located somewhere between Lan and Wan.

Public. The user has the right and get this service.

Private. To access it you need to have an Access Key so one must be authorized personnel.

Internet. It is a worldwide public network interconnected through routers across the world, servers with special features: voice, data and video, allows for feedback (Acknowledgement = ACK) is interactive. Applications run on the server. Among its features are:

Protocol TCP / IP communication. It is thanks to graphic browser and has a protocol for transmission: Hypertext Markup Language, HTML. Use auxiliary languages (that can handle structures: If, for and while) for managing database as HTML is unstructured language. These languages can be Java, ASP, PHP, PER, etc. Using network operating systems, NOS. Internet problems are related to the following three aspects to the extent that they are considering the use of these networks become more valuable:

• Security. Internet is born without security.

• Dependence of networks. It is a distributed network.

• Administration. It requires specialized personnel.

Origin of Internet. The Department of Defense, DOD. EU wanted to develop a global network for military purposes. The Advanced Research Projects Agency, ARPA. In 1965 ARPANET think that is a program of research that should develop techniques and technologies to connect different types of networks and communication protocols that allow computers connected freely communicate across different platforms and networks. They managed to communicate two hosts and later in 1969 in conjunction with scientists from the University STANFORD achieved the set with more than 100 TCP / IP protocols, this was rejected by the industry standard through international organizations. The US Army declare a standard in 1983. In 1985 was adopted by the National Science Foundation, NSF of United States, developed a network called the Internet, non-profit and international several universities in the world, was regulate by the InterNIC for administration and authority for allocation of Internet numbers, IANA, Internet assigned number authority was created. In 1989, a Netscape scientist invents browser with HTTP and HTML making the most friendly, graphical and easy network. In 1991-1992, the Internet is released by the US to the world and that's when exponential growth begins and later the management of Internet at 1993 IAB Internet Architecture Board administrate with two organization: IETF (Internet engineering tag force) and IRTF (Internet research tag force).

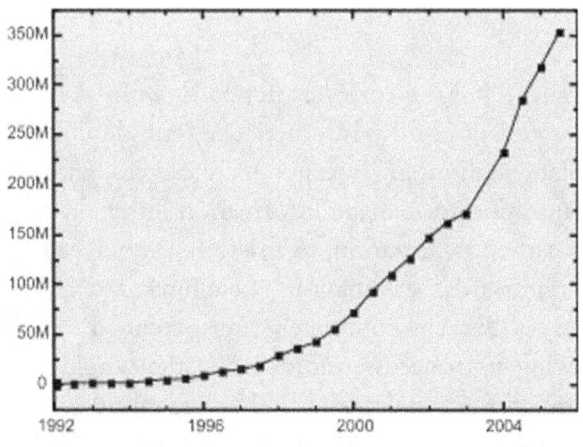

Figure 3. Millions of Internet Users / Years
(Source: ISC)

In the Figure 3 the vertical scale indicates million Internet users, the horizontal years. By 2006 there were about 350 million users.

To connect to the Internet must go to a provider of Internet services ISP

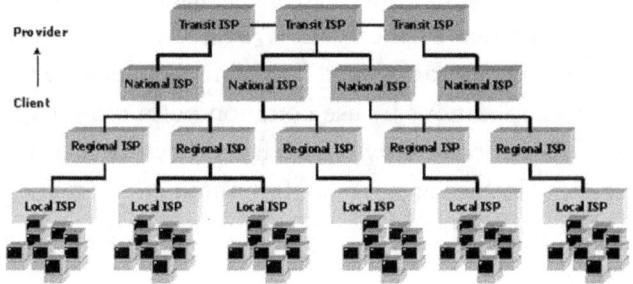

Figure 4. ISP Levels

Intranet, It is a private network with the same characteristics of the Internet. The term describes the implementation of Internet technologies within an organization to manage information internally to the company (Organization, newsletters, employee of the month awards, terminations, new hires, etc.). This is done so that it is completely transparent to the user, allowing it to access, individually, the whole set of information resources of the organization, with minimal cost, time and effort. Members of the same use, as is presumable, web clients to access the information. Will be implemented, therefore, TCP / IP, and for creating HTML documents is used. The disparate platforms and information systems in an organization, and issues to share information between them, forcing those responsible for information systems integration solutions look, reliable results and an acceptable cost.

Using the World Wide Web, technology for its ease of implementation, low cost, and quick apprehension and acceptance by the user, as well as its portability to different platforms, and its ability to interact with various applications by use common output interface, CGI (Common Gateway interface). Factors that are heavily influencing takeoff Intranet can be summarized as follows:

• Easy configuration and adaptation to the technological infrastructure of the organization and management and manipulation.

• Affordable Cost.

• Adapting to the needs of different levels: company, department, business area.

• Easy integration of multimedia.

• Available on all computer platforms.

• Possibility of integration with internal databases of the organization data.

• Fast training.

• Access to the Internet, both abroad and domestically, by register network users with access control.

• Use of public and open standards, independent of outside companies, as may be TCP / IP or HTML.

Extranet, It is accessible from anywhere in the world via the Internet to a private network. It behaves like a "tunnel" in the Internet.

Three basic characteristics of an extranet:

1. The need to encrypt data. Involves applying a function to some data so that these remain unrecognizable. The security of an encryption method

is measure network in terms of two things: the number of possible keys and would have to try to decrypt the password and complexity of this calculation, which affects how many checks can be made per second. The number of possible keys is usually the most important factor, and is typically measure network in the number of bits occupied by each key.

2. Need for an access code is an alphanumeric string to be dialed to get a service, benefit or to achieve

3. Integrity of information. Let this be useful and stay integrated. Extranet creates a common means of communication for groups (customers, agents, suppliers, etc.) closest to the organization and differential treatment with both end users and employees. Extranet applications of Internet technology for working in real time.

B) Architectures Protocols = Logical Topologies:

It is a formal set of agreements and rules that govern how computers should communicate over networks by minimizing transmission errors. These carry the fragmented information, so no transmission, however large, monopolizing network services.

A protocol describes:

• Time on the exchange of messages between two communication systems.

• The format of the message must have for sharing between two computers that use different protocols can be established.

• What actions to take in case of errors.

• The assumptions made about the environment in which the protocol will be implemented.

Classification of protocols:

According to Level:

High. TCP / IP (Owner of universal application), OSI MODEL (from the industry) and SNA (system network architecture, IBM owner).

Lower. Any level lower than application. Examples: 802.1, 802.2, ... 802.14

According to Access:

Connection less: No acknowledgment (ACK).

Connection oriented. Virtual Circuit. No recognition.

According to the manufacturer:

Industry: These are standards set by international

organizations.

Owner: are owned by a particular manufacturer and are used universally.

Circuit: It's very similar to a telephone conversation; three steps must be 1 Application, 2 Data Flow and 3 Termination.

TCP / IP (version 4) is by classification own manufacturer, it is owned by the DOD. IP V4 has 32 bit is not standardized. IP version 6 has 128 Bit is the industry standard, SNA proprietary also as part of the IBM company. IEEE standards low. (80 because they met in 1980 and 2 February). Here are some examples:

802.2 LLC and MAC.

802.3 CSMA / CD.

802.4 Token-passing.

802.5 Token Ring..

802.6 Man.

802.7 Broadband.

802.8 used by the Technical Advisory Group fiber.

802.9 Define network integrated V-D-V.

802.10 security.

802.11 wireless.

802.12 Access Tecnologys.

802.13 unused.

802.14 defines the standard cable modem.

802.15 WPAN.

802.16 W broadband.

IEEE. US Agency, part of ANSI, which promotes research by own rules of standardization. One of its main activities is the development of non-binding rules, but generally accepted in the area of communications and electronics, with emphasis on measurement techniques and definitions of terms.

Topology: The shape adopted a schematic wiring diagram or physical layout of the network devices to link to a share network medium.

Types of Topologies:

Bus. It consists of a single cable which will connect all the workstations. In this system a single computer at a time can send data which are heard by all the computers that make up the bus, but only the designated uses. Advantages: The wiring is easy to implement and inexpensive. Disadvantages. If you have too many computers connected simultaneously, low efficiency greatly. Use coaxial cable and its speed is 10 base 5 (10 = bandwidth, signaling and type base = 5 = thick coaxial) or 10 base 2 depending on rope thickness and distance. A cut anywhere breaks the cable connection nodes of the entire segment.

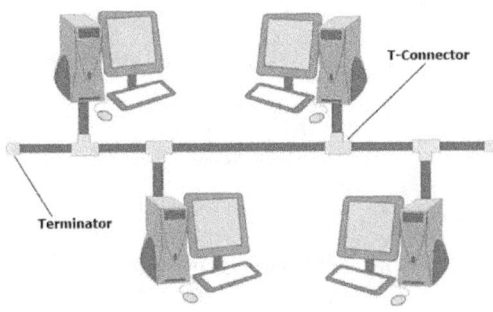

Figure 5. Bus Topology

Ring topology. It is a development that IBM is to connect each station to another to form a ring. Servers can be anywhere in the ring and the information is passed in one direction from one to another station until it reaches its

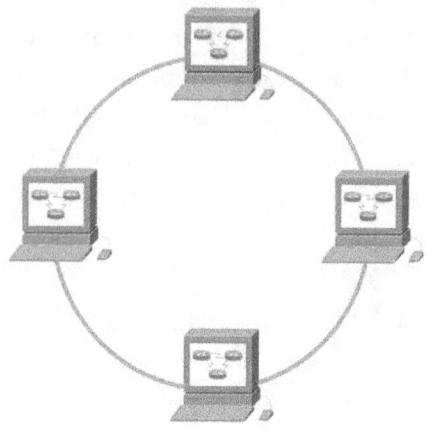

Figure 6. Ring Topology

Disadvantage. The failure of a station disrupts the entire network. Currently there are no physical connections between stations, but exist central wiring or MAU that implements the logic ring. Costs could be.

Advantage. With distribution by optical fiber cable (FDDI) has double ring network redundancy.

Star topology. In this scheme where all stations are connected to a hub or computer HUB cable. For future extensions may be placed other HUBs cascade resulting in the hierarchical star.

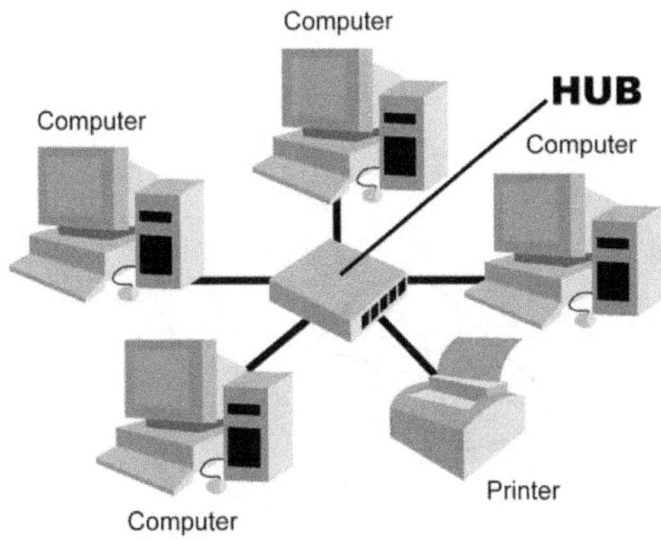

Figure 7. Star Topology

Disadvantage. Everything is centralized and wiring is complex and costly. Advantage. Easy management and scalable good speeds.

Mesh topology. Here all nodes are interconnected by a full mesh. Allowing Network redundancy between nodes, that is, if a route fails the other routes are available which by the formula (r = Network redundancy links and node= numbers of points) follow:

$$r = \frac{n(n-1)}{2}$$

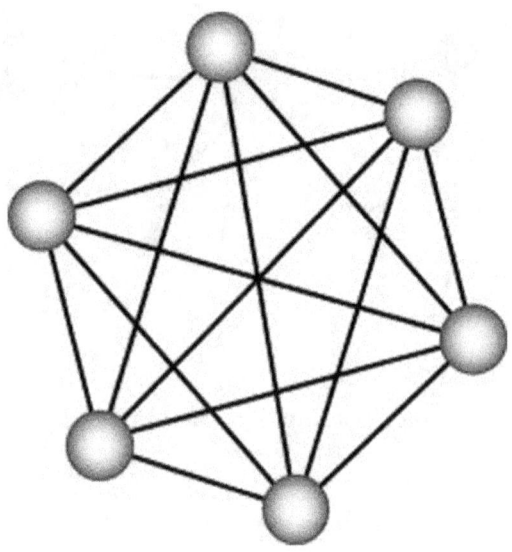

Figure 8. Mesh topology

What all the above topologies are that we should use?

In modern networks are a combination of all, but the most common is the star topology and from this point of view we can design our LAN.

Example: If we connect in chain 6 concentrator (switches) 12-port, low processing relatively inexpensive. People who do not know the cascade using the ports for PC, this causes inconsistency (in an instant can only have about 36 ports operating simultaneously of the total 60, the LAN would have intermittent service) because the latency, is recommended connect up to 3 hubs to have 4 network segment, 3x4 (cannot connect infinitely equipment), in other words, the design should be a star. To correct the problem we select one of 6, and use as central and feed all from this, now in star topology no exist the problem of inconsistency. This error is very common. As we take a 12-port hub with the same performance of the other two things would improve: we must first change the core team for team Backbone and get more processing and more expensive to increase performance and second would have to put a second equipment to get more processing and backup.

2. OSI MODEL, OPEN SYSTEM INTERCONNECTION

Why is it important to study the OSI Model?

The OSI model is important because it Networks complexity standardizes interfaces, allows modular engineering, ensures interoperability of technologies, accelerates development and simplifies the teaching and learning. Of these six reasons for all these features the most important is the integration of different technologies. This protocol is used as a reference in the industry and to the extent that we know we can talk relate consistently with people in the area. We memorize the number and name of the level.

All layers are hardware and software components being most abundant in the upper layers software and hardware equivalently is more substantial in the lower layers. Layers 1 to 4 are called internetworking and 3 uppers layer are interoperability.

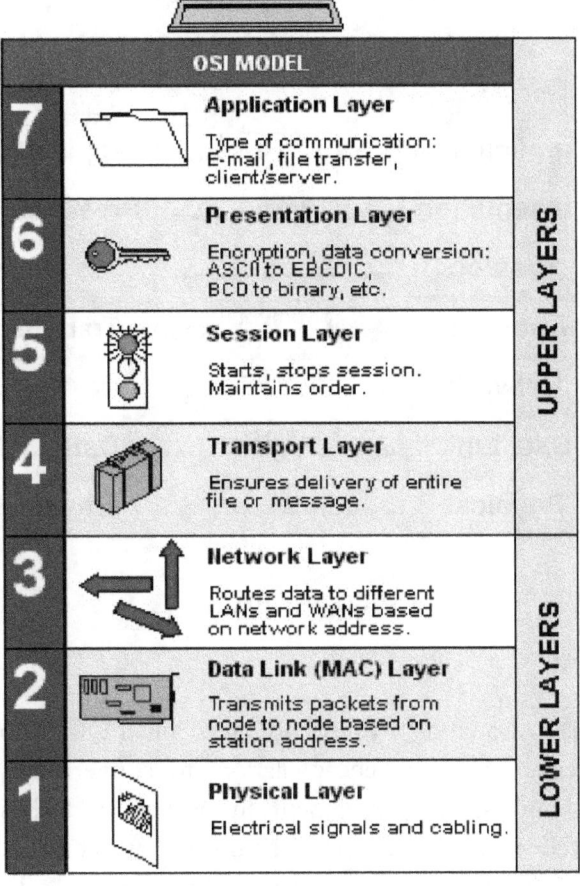

Figure 9. OSI Model

= Integration + Interconnectivity Internetworking.

OSI help to understand how the data is place. Is moved through the OSI model and is very simple. The data according to the level have a name.

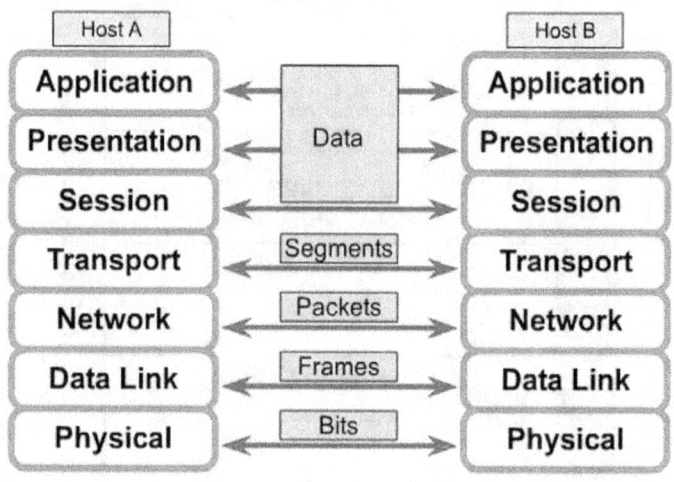

Figure 10. PDU

This is known as a Protocol Data Unit, PDU. Thus we can define the encapsulation; to convert these data into segment, segment in packets and these packets into frames for transmission at bit in the media (data-segment-pack-frames-bit). On the other hand it is received at the other end defining desencapsulation; which converts these bits into frames-packets-segments-data so that the original data is obtained without heading. Necessarily each end must have 7 levels.

Defining OSI Levels:

Level 7- Application. Tools for easy use by the user. Email management, FTP (file transfer protocol), Internet browsing, HTTP (Hypertext Transmission Protocol), Telnet remote access (Terminal Network) and others.

Level 6- Presentation. Reorganization of the data. For example Handlers Database, DBMS (Data Base Management System); SQL Server (Structure Query language server), Oracol among others. They are software interfaces between the user and the equipment.

Level 5- Session. Time control in the dialogue between applications. Use difficult applications such as database standard Structure Query Language, SQL.

Level 4- Transportation. Congestion, windowing and buffering. Here the flow control and the Integrity of the data that will be determined by the balance of three elements defined. This level also is responsible for correcting errors in the encapsulation.

Level 3 network. The dominant team at this level is Rauter its main function is to route IP packets (versions 4-6) or logical addresses. It is important to know the class hierarchy, sub netting (subnets) and decimal system.

Level 2 Link. It is the only level with two Sub-levels LLC (Logical Link Control) and MAC (Media Access Control) addresses hardware. Here prevails the network card and Hub (swicht). It is important to know the hexadecimal system. Also this level is responsible for the notification of errors in the encapsulation.

Level 1 - Physical. The medium may be determined by the cable or not cable (Wireless). Hub, modem and multipexores are in this level. Also repeaters extend the length of the network connecting two segments and amplifying the signal, but along with it also amplify the noise. It is important to know the power, attenuation, synchronism, symmetry, analog and digital systems.

OSI MODEL EXAMPLES			
Level	Level Name	Protocols	Equipment Technologies

7	Aplication	SNMP DNS FTP TFTP Telnet SNMP SMTP TFTP	
6	Presentation	JPG DBMS PER JAVA ASP	
5	Session	SQL	
4	Transport	TCP UDP SPX	
3	Network	IP IPX Apple TAlk	ROUTER

| 2 | Link | Mac LLC | SWICHT MPLS ATM |
| 1 | Physical | Ethernet | HUB Multiplexor Modem |

3. PHYSICAL LEVEL

As seen above the OSI model offers advantages to be exploited in this chapter and the following to understand data networks. In that order we begin with level one and below.

The Physical layer defines the standards and protocols used in the connection. It also defines the cables and connectors. At this level are the Hub, repeaters, multiplexers, modems and more. This electric current is the movement of electrons in a conductive material such as copper. There are two types of current: direct and alternating AC and DC. The DC current is found in car batteries, can also produce static when rubbing two elements, it is found in nature. Instead AC or induced current is produced by electromagnetic induction (Miguel Faraday), it is simply an iron core wrapped with copper wire next to a moving magnet. The latter can see it in our home where we connect the appliances, they have three contacts: The ground for protection of persons and equipment, the shorter is the hot line or line alive and length is the return (neutral) which closes the circuit.

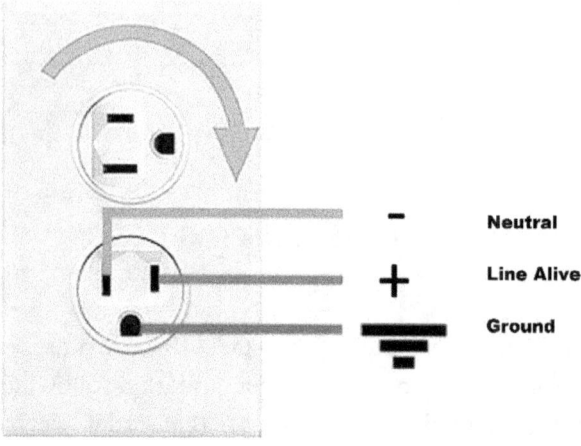

Figure 11. Alternal Current

The transmitters and receivers use AC but internally have a transformer to reduce it and a diode bridge to convert it to DC. Mr. Coulomb studied the behavior of loads and found that opposite charges attract while like charges repel. The diode has two electrodes; one anode is positive and a negative that is the cathode. The main characteristic of the diode is that the current does not pass in the reverse direction, so if we apply an alternating current between the plate (anode) and the cathode plate with respect to the cathode is sometimes positive and sometimes negative other thus will vary depending on the current cycle but always in the same direction.

If a diode passes a current such as that shown at the

top of the figure, flows a stream as shown in the lower part thereof. This action is called correction that results in a polarity to make it straight (DC current) is added a capacitor for maintaining peak. The following figure is a diode bridge:

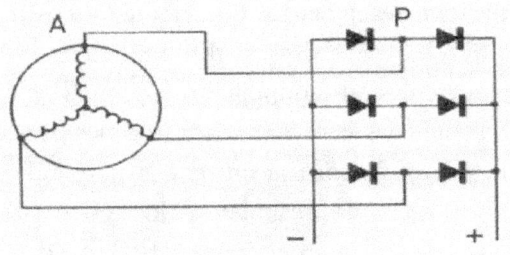

Figure 12. Rectifier current

Attenuation:

Is the signal loss in the medium caused by the distance or noise. The measure by electrical impedance:

$$Z=R+Ri+Rc$$

R, resistance depends on the thickness of the wire, which can compare with a water hose while it is larger allows more water to pass into the most common case. Ri, inductive reactance lost, this can easily be seen in transformers (small and large coils that when combined down or up the AC current) that are on these power lines should increase current to carry it further (This solve the problem of long distance) and then tape to use it. The losses generated by these devices are notable for the amount of heat generated. Rc, capacitive reactance or capacitance loss. In power lines when current passes through them forms a reverse current in the opposite direction (Lenz's Law) like Newton law, every action a reaction, these two opposing currents form a current is perpendicular to these, form a circular magnetic field to the wire when it hits the ground in an electromagnetic field resulting from a capacitor (Two parallel metal plates one + and the other - that can store current) throughout the course of the wire, these losses are significant, we call noise . In data telecommunications this noise caused by the capacitive reactance remove it by twisting the wire to cancel the electric field.

Synchronous Transmission. Requires transmission of both data and a clock signal to mark the compass sent to synchronize the transmitter and receiver. This synchronization is achieved with a master and a slave or internal clock or external clock.

Asynchronous Transmission. The synchronization process between sender and receiver is performed on each transmitted code word, start and end flag (Flag). This is accomplished through special bits that help define the environment of each code. Imagine that the transmission line is at rest when it has the logic "1". One way to inform the receiver that is going to get a character that character is putting one start bit, "start bit" with the logic "0". Once we receive all data bits are added to one or more stop bits, "stop bits" logic level "1" to reset to the initial state of the data line, leaving it ready for transmission of the next character. For example, if one considers a system of asynchronous transmission with 1 start bit, 8 information bits per code word and 2 stop bits, burst transfer have 11 bits for each transmitted character. A lack of synchrony affect at most 11 bits, but the arrival of the next character, with his new start bit, cause a resynchronization of the transmission process.

Analog signal. The analog signals are continuous wave, for example, our voice on the phone and passes this retransmitted waves which are analogous. Herls found that these waves travel through the air (via the ehter erroneously as this does not exist, simply move alone in the air) and in his honor can measure them at a rate of seconds. The voice can reach up to 4 KH (H=1/s)so the voice can oscillate 4000 times a second.

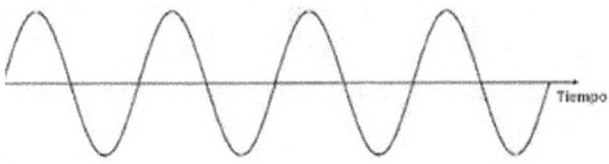

Figure 13. Analog Signal

Digital signal. Changes Between two states, namely: (1) Presence or Absence (0) voltage.

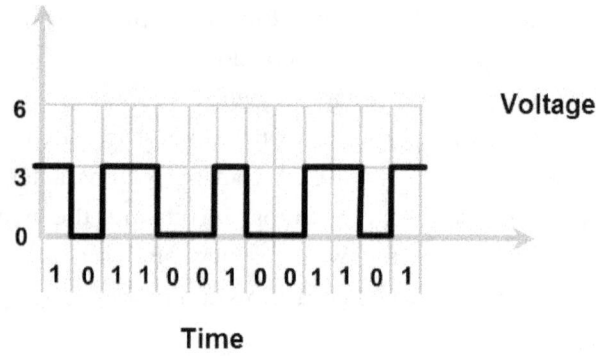

Figure 14: Digital Signal

Symmetry. When you split an item into two parts are homogeneous. From Asymmetry to an element in two parts are mixed. Modem. (Modulator - Demodulator) is a device that converts analog signals to digital and vice versa. According to the Federal Communication Commission, FCC speed up (upstream) is 33.3 and down (down streams) is 53.3. These equipment are asymmetrical and asynchronous. The modem receives the digital signal and transmitted on the same medium that phone use, this analogy can reach up to seven kilometers without repeaters. The problem with this system is that when wearing a longer distance also amplifies noise, 200 kilometers, about 20 repeaters would be needed at the end for more filters we use, come only noise.

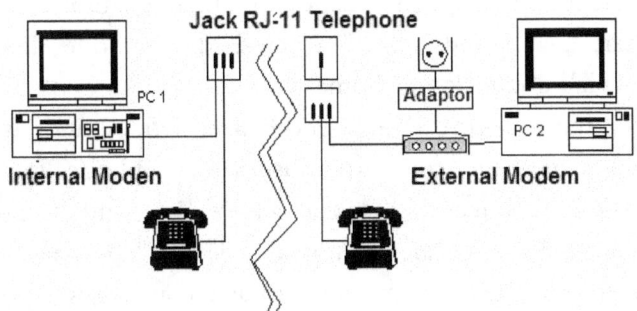

Figure 15: Modem

The v.92 protocol is a new specification for modems developed by the International Telecommunication Union, ITU. Advantages of the Protocol against the protocol v.92 and v.90: In v.92 data compression 4: 1-6: 1. Connections up to 50% faster. Upstream up to 48Kbps. But the most striking feature is Modem-on-Hold, you can receive calls without disconnecting the Internet line, although it is necessary to have contracted the call waiting Telephone Company; you can even make voice calls without disconnecting the data line. Technology v.92, like previous v.90 and v.34, must be present on both the client computer and the server, so it is necessary that your provider has updated the access nodes to support this technology. The waiting call without dropping the connection will work with any supplier at the end of the line available nodes compatible with v.92.

ADSL Modem. Asymmetric Digital Subscriber Line. Under the name of a series of xDSL technologies that enable the use of a standard telephone line (which connects our home with Telephonic Central) for transmission of high speed data defined and, at the same time, for normal use as a telephone line. ADSL is a technology that allows baseband using standard phone lines to transmit data at high speed, with constant access and simultaneously use the phone to talk. Separate voice that sends data on different frequencies, so that you can talk on the phone while the computer is connected to Internet.

Multiplexer. The multiplexer is basically multiple inputs and a single output that lets you select the channel you want. This can receive analog or digital signal, it is important to always transmit digitally in the middle, there is no noise when the signal is amplified because the 0s and 1s are regenerated. The distance without repeaters can reach a kilometer by copper in reference, microwave about 70 kil and fiber about 100 kil regardless of the medium is solved the noise problem. Passes the signal from analog to digital and vice versa. Save lines on the street. Lower cost. The signal from your phone (analog), passes the (digital) station, then the (analog) system to digital and then finally another analog phone.

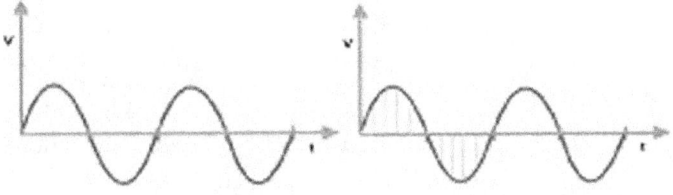

Figure 16. Analog signal samples

Nyquist theorem. An analog signal can be reconstructed without error with samples taken at equal time intervals. The sampling rate is equal to or greater than twice the frequency (4-8) of the analog signal (4 kHz) . He showed that in a second you can send 64,000 bits, which is called digital signal zero DS0 (Digital Signal 0). Considering that the voice has an average frequency of 4 KHz, sent found that the optimum is 8 bit (sample) x 8 (2f) = 64 Kb / sec.

TDM, time division multiplexing. Time division multiplexing. Technique which can allocate bandwidth to the information from multiple channels on a single cable, based on reassigned time slots. Each channel bandwidth is allocated, regardless of whether the station has data to transmit. Example:

PCM, Pulse code modulation. Analog transmission of information in digital form by sampling and sample coding with a fixed number of bits.

24 channel multiplexer (PCM). To calculate the time. Each channel can send 8 samples (remember the Nyquist theorem) Moves spinning at full speed without the user being aware of it. Therefore, sends 24 x 8 = 192 but should added one to indicate that the cycle starts, is the bit that controls the time synchronism. Hence the correct calculation is:

24 x 8 = 192 + 1 (sinc) = 193

If you have 24 channels and sends 193 bit, the time delay will be:

24 / 193 = 0.125 milliseconds

Invisible to us humans. To find the resultant velocity:

64k (DS0) x 24 channels = 1,536,000(Data)

$$+ \qquad 8,000(Sin.)$$

1.544   Mb/seg   =   T1
(Data+Sinc)

T1 is an American standard ..
In Europe is called E1 and is equal to
32 x 64 = 2.084 Mb / sec (use two
channels of 64 for synchronization)

Thanks to the multiplexer we can have trunks between PBXs, otherwise assuming these were we could not see the sky for the amount of cables.

Examples: Telephone companies join call centers and Internet connections with multiplexer from 12 T3 (45 Mbit) =T5 (45Mbit-8000 people simultaneously talking on a single cable) this system is also called STM1 and can grow in the following manner: STM4= 48 T3, STM16= 192 T3, STM64 = 768 y STM 256 (STM1, Synchronous Transport Module level 1). Multiplexing has problems when samples are consecutive zeros because no voltage is zero, the synchronism is lost. To fix this there are two algorithms AMI and B8ZS:

Investment alternate brand AMI. Line code type used on T1 and E1 circuits. Making the last and each sample placed on a heart 1, to maintain synchronization. You cannot use all the bandwidth. 56 kb / s for data and 8 kb / s for sync.

Binary 8 zero substitution, B8ZS. Use a bipolar signal has values + and -. You can use the full bandwidth, so it is called Clear Channel.

There are more forms of multiplexing: FDM Frequency Division Multiplexing. Technique by which you can assign bandwidth to information from multiple channels on a single cable, based on frequency and among many others, the TDM was analyzed because it is the most used.

Structure cabling: The issue is extensive, however due to its great importance, here a brief description using star topology and pair wire, UTP (unshielded twisted pair), 100 Base TX. As shown in Figure most important of this wire braids are removed the noise.

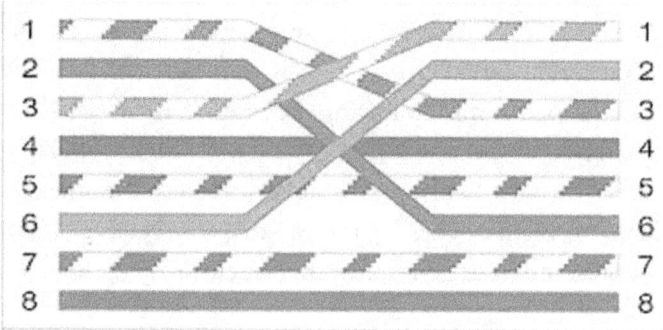

Figure 17: Twisted Pair UTP

Example: Let's analyze a structure cabling system to star with 100 computers distributed in 4 classrooms. Doing a simple we need to apply rule of 3X4, we can put on any case in reality. The first thing to consider is the placement of the equipment room should be located according to the rule of measure, it should preferably be in a central position, here the racks are installed, the hub (witch, equipment assets) and patch panels. For connections, the pipes that carry the UTP cable to each computer, we use 4-pipe 3 inch c / u to 25 computers in each classroom. The number of tubes depends on the distance of each classroom, which should not exceed 100 meters. The amount of cable to be used is determined by a factor of 0.17, based on the experience of such facilities. In this case is 0.17x100 = 17 cable boxes 1000 feet each. The design of the 100 points is represented by the following Figure17.

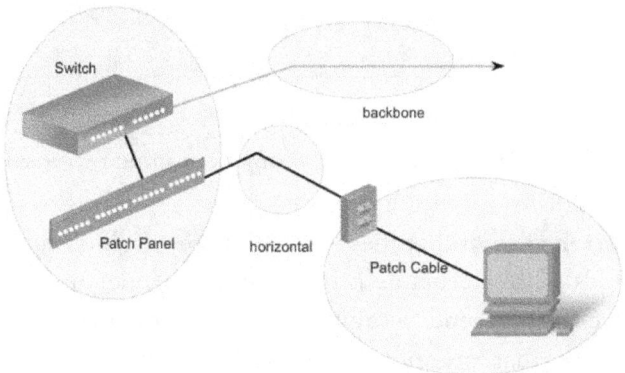

Figure 18: Horizontal Cable

The horizontal cable from the connector (RJ45 Jack) to patch panel is 90 meters direct connection should be according to the following Figure 19 on the ends to complete the 5 + 5 = 10 meters for a total of 100 meters.

Structure cabling is defined by the sum of the horizontal and vertical wiring or cabling backbone.

Economic proposal horizontal cabling network 100 points:

Cantidad	No. Parte	Descripción	Precio	Total RD$
		Cableado Horizontal 100 Puntos de Red Categoria 5		
17		Cajas de cable UTP mil pies	4,320.00	73,440.00
100		Pach cord 7 pies	138.00	13,800.00
100		Pach cord 3 pies	96.00	9,600.00
100		Jack RJ45	168.00	16,800.00
100		Cajas de pared	72.00	7,200.00
100		Face plate	36.00	3,600.00
5		Pach pannel 24 ptos cat 5	4,080.00	20,400.00
5		organizador de cable	9,000.00	45,000.00
1		rack 7 pies	9,000.00	9,000.00
100		Mano de obra instalacion	780.00	78,000.00
			Subtotal	276,840.00
			ITBIS	33,220.80
			TOTAL RD$	310,060.80

Figure 19: Horizontal Cabling economic proposal

To the extent that these variables are weighted can win bids with acceptable profit margins. Under no reason we recommend working with inferior materials or those that are not recognized by the international standards.

Financial proposal for vertical cabling GB 100 network points: Here the profit margin is 15%. Vertical cabling should be Category 6 UTP or upper.

Cantidad	No. Parte	Descripción	Precio	Total RD$
		CableadoVertical UTP GB 100 Puntos de Red		
5		Switch 24 puertos 10/100 y 2 puertos 10/100/1000	3,448.00	17,240.00
4		Pach cord UTP Categoria 6 Gbit 7 Pies Conexion crusada	600.00	2,400.00
			Subtotal	19,640.00
			ITBIS	2,356.80
			TOTAL RD$	21,996.80

Figure 20: Vertical Cable or Backbone

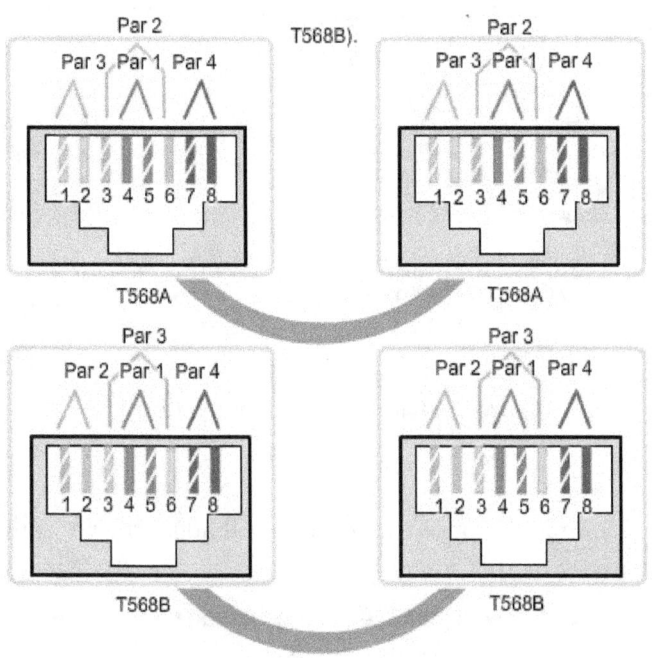

Figure 21: Direct Cable Connection

The following Figure the cross connection for simple connections and 1Gbps expressed.

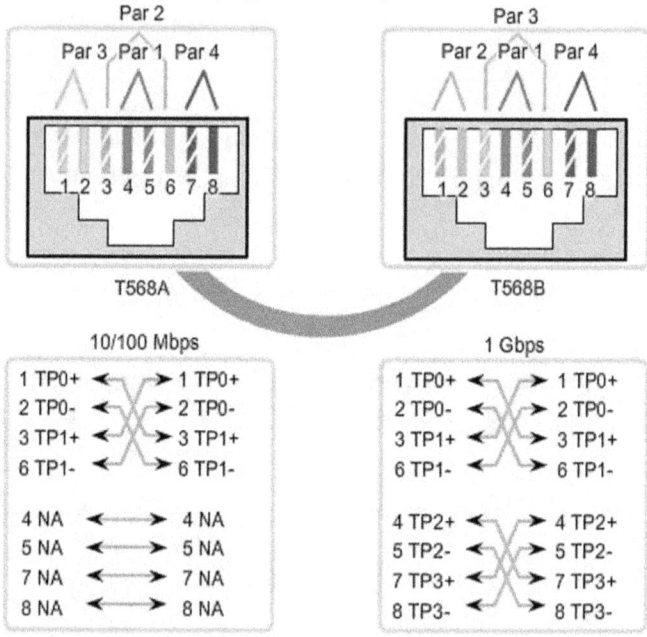

Figure 22: Cable Cross Connection

Economic proposal vertical cabling fiber optic network to 100 points:

Cantidad	No. Parte	Descripción	Precio	Total RD$
		CableadoVertical Fibra Optica GB 100 Puntos de Red		
5		Switch 24 puertos 10/100 y 2 puertos Fx	4,500.00	22,500.00
4		Pach cord Fibra Optica 7 Pies Conexion crusada	1,600.00	6,400.00
			Subtotal	28,900.00
			ITBIS	3,468.00
			TOTAL RD$	32,368.00

Figure 23: Vertical Cable or Backbone

These two proposals have the same vertical cabling performance. Using Gbps UTP we suggested to switch that are all places very close. The use of optical fiber is recommended for example in a different building floor.

4. LEVEL LINK

The link layer-sublayers: the Media Access Control (MAC) and Logical Link Control (LLC). Bridges (bridges) and SW operate in the MAC layer. The function of layer two is to ensure the transfer of error-free data between nodes, also establishes the medium access control.

The upper Level 3 handles software addresses or logic address so to communicate with level two requires the LLC sub layer because the direction of level 2 is hardware and cannot communicate directly.. For example, the ARP / RARP protocol are what make address resolution TCP / IP.

802.2 Logical Link Control Layer				2 Data Link Layer	LLC
					MAC SAP
802.11 Medium Access Control Layer					MAC
					PLCP SAP
					PLCP
IR	FHSS	DSSS	OFDM	1 Physical Layer	PMD SAP
					PMD

Figure 24. LLC

LLC Logical Link Control IEEE 802.2 and Address Resolution Protocol, ARP This protocol is trigger when unknown MAC address. When a network node wants to send data to another node, you must find your physical address. The source node knows its own IP address and physical address (it is its network interface), but all I know the remote computer's IP address. For the equivalent physical address, and ARP (broadcast) is sent. This message is received all computers on the same physical network, but only answers the requesting computer. ARP is responsible for resolving IP addresses into physical addresses.

RARP. Reverse Address Resolution Protocol. This protocol is activated when the IP address is unknown. In some situations the remote computer does not know its IP address, that is, in the case of diskless stations, the only difference between these is the physical address of the network interface. Because you need to communicate with other IP address, sends a RARP (Broadcast) package. This packet is received by a server that contains a table of resolution between physical addresses and IP addresses. This server resolves the IP address and sends it touches you

Dynamic Host Protocol, Configuring DHCP. Use RARP. It is a network service used to assign a dynamic

IP, it allows to give an IP address to each client network, as it connects. Every machine on a network using TCP / IP requires a unique IP address, it is not repeated in the network. For this, there is a possibility that the network administrator, attributed an IP manually to each machine and static, which often involves a tedious job, or uses a DHCP server, which is what will take care of perform this task when a machine connects to the network. Medium Access Control, MAC address is a hardware call that is printed on the NIC (Network Interface Card). This is 48 expressed in binary bits. Shown in the 12-digit Hexadecimal. The first half of the IEEE assigned manufacturer (left side) and the second half is the serial.

The NIC is at Level 2 in the OSI reference model. Since the TCP / IP protocol unifies his Level 1 and 2 1 OSI, TCP / IP NIC is level 1. Generally speaking it is implied that 2 is in relation to the OSI model.

Telecommunications should be analyzed from the point of view of LAN and WAN and independently for better understanding:

LAN, It is characterized by low latency:

IEEE 802.3   - Ethernet (CSMA/CD).

IEEE 802.5   - Token Ring.

ANSI FDDI - Token Ring (Fibra).

WAN, It is characterized by high latency:

PPP

HDLC

FRAME RELAY

MPLS

RS-232   Asymmetric Ports (Modem)

V.35   Synchronous Ports (Multiplexers)

Cyclical redundan Check, CRC: Algorithms for detection of known errors. It is a way to check errors in a message by mathematical calculation with the number of bits in the message. This number is sent to the recipient with the information that proves that you have received and repeat the math. If there is any difference between the two calculations, the receiver requests to be send the information again.

Frame Check Sequence, FCS. This algorithm also provides a mechanism for error detection in case of data corruption.

Broadcast: Type of communication in which all potential recipient is reached by a single transmission. All share the same medium. When there is much flooding, there are many collisions.

Broadcast :

• Flooding. When a message is sent to everyone and everyone picks it up.

• Unicast. In unicast transmission system information is transmitted aimed at a single point. An address that can only be answer by a single host.

• Multicast. In the multicast transmission system information is transmitted to a group. An address that can only be answer by a group of Host.

Segmentation. Is used to solve the problem of creating broadcast collision domains smaller (micro segmentation). The Hub (Level 1), Switch and Bridge (Level 2), work on a broadcast domain. But the Switch and Bridge know who is in the broadcast domain because they create a Mac Address table and divide it into collision domains. The number of collision domains depends on the number of ports.

The router knows all that is outside the domain of Broadcast. Have many ports, many broadcast domains and collision .

Arbitration. As it is a share medium equipment must have controlled access. Here we only analyze the IEEE 802.3 and 802.5.

802.5. Token Ring. Solve the problem of the share arbitrator putting the Token, this algorithm is deterministic, only one device can transmit at a given time.

802.3. Ethernet. To solve the problem of share medium in this system should reduce collisions. Arbitration is performed with algorithms Back Off and CSMA / CD.

Collisions. It is assumed that each bit remains in the domain maximum time "time slot" of 25.6 ms (more than 25 millionths of a second), which means that at this time must have reached the end of the segment.

If at this time the signal is not out of the segment, it may happen that a second station on the part of the segment not yet reached by the signal, it may begin transmitting, since the carrier sense indicates that the line is clear, given that the first signal has not yet reached the second station. In this circumstance MA multiple access and collision of both datagrams is inevitable happens. In the operation of an Ethernet network is normal a certain rate of collisions, but should be kept as low as possible. In this regard a normal network must have less than 1% of the total collision of packets (preferably below 0.5%). To perform this type of testing is necessary to have adequate analyzers.

BackOff. Time it takes the computer to retransmit after a collision. This time is 0.52 milliseconds.

Collision detection sensor carrier multiple access, CSMA / CD (Carrier Sense Multiple Access /Coalition Detection). It is used to avoid collisions. Assigns a delay and makes a drawing for each, assigning a random number using an algorithm that accounts for the differential. In full duplex, CSMA / CD is disabled. The CSMA / CD protocol used in Ethernet. It is based on when a Data Terminal equipment, DTE connected to a LAN wants to transmit, it stays on the line until no team is transmitting (CS is the "Carrier Sense" part of the protocol); once the network is quiet, the computer sends the first packet. The fact that any DTE can gain access to the network is the MA "Multple Access" part of the protocol. From this moment comes in part collision detection, CD (Collision Detection), which is responsible for verifying that the packets have reached their destination without colliding with those who may have been sent by mistake by other stations. In a collision, the DTEs detect and suspend the transmission; Each DTE waits a certain period, pseudo random, before resuming transmission.

Full-Duplex. This term indicates the characteristic of any hardware, to send and receive data at the same time through the same means.

Half-Duplex. It only allows the signal in one direction at a same time by the same media. For example, mobile radio systems.

Switching. Basic equipment for LAN networks which operates at Layer 2 of the OSI model. Its processor is hardware base while bridges are software base, hence the switch is often refer as a multiport bridge, but with a lower cost, higher performance and increased port density.

The switch at Layer 2 makes its data forwarding decisions based on the destination MAC address contained in each frame. These, like the bridges on the network segment collision domains, providing greater bandwidth for each station.

A switch can have four states and happens during the startup process:

1. Blocking
2. Listening
3. Learning
4. Forwarding

From Step 1 through 2 takes 20 seconds in the 2 to 3, delay 15 sec and 3 to 4 15 sec. 50 seconds total delay from start until it reaches the optimum operating time is approximately one minute.

When a Switch turn on the current memory is not activated yet, everything comes: it call flooding, after the macaddress is learn in its hardware address table. The minute after convergence known devices connected to its ports can send the signal (frame) to the address indicated in the form of unicast.

Loops Avoidance, Correction of duplicity. In the Switch loops endless form. There is an algorithm called Spanning Tree, which is used to eliminate loops and leaving trails as redundancy for backup. The IEEE 802.1D Spanning Tree Bridge and Switch eliminate duplicate paths and links in a network. The protocol allows the switch to communicate with these other devices and over the network.

Forwarding: Process by which an Ethernet switch or bridge reads the contents of a packet and transmits it to the appropriate segment. The remission rate is the time it takes the device to perform all these steps. Spanning Tree Protocol: In a topology witch Connecting 3 Switch.

1. If the switch has 24 ports only have a Mac - Address that represents all ports. STP select the better bandwidth ID or the one with the smallest MAC, this Switch becomes root and put all ports in a locked

state.

2. For each port the Switch connected a segment.

3. A non-root changes it locked to open (forwarding).

4. The ports are closed, leave it as backup.

Spanning Tree haves:

- 1 root per system
- 1 port per segment
- 1 port per non-root

Convergence. It is the time to passes from the blocked state to Forwarding state and the algorithm Spanning Tree (STP) has been activated. (About a minute).

Filtering Mode Cut-through: The technique to examine incoming packets for which an Ethernet switch looks at the MAC of a frame and forwards only (Fast Filter). This process is faster than looking at the whole package, but also allows some packets to transmit error. For this system latency is the same

Filtering Mode Store and Forward. The technique to examine incoming packets for which an Ethernet switch analyzes the frame and forwards (Slow Filtering). This process is slower and does not transmit frame with errors. For this system latency varies according to the size of the packets.

5. NETWORK LEVEL

Routers operate at layer 3 of the OSI model and therefore distinguish and routing decisions based on the different protocols of the network layer. Routers placed boundaries between network segments because they send only traffic that is directed at them, eliminating the possibility of broadcasts, transmission of packets of unsupported protocols and transmission of packets destined for unknown networks.

To accomplish these tasks, an Rauter performs two functions: Create and maintain a routing table for each protocol in the network layer. This table can be created manually using static routing or by dynamically route protocols (RIP, OSPF, etc.)

Identify the protocol contained in each package, extract the destination address of the network layer and send the data based on the routing decision.Routers select the best path to send data based on metrics (# of hops, speed, transmission cost, delay and traffic conditions) routes. Additionally, they have the ability to implement security policies and bandwidth utilization. But, on the contrary, the process to be performed with the packet is reflected in increased latency and decreased performance.

Features:

• It connects networks using different identities.
• Only transmits the data needed by the final destination through the network.
• examines and reconstructs packets bypassing the errors to the next network.
• A Rauter stores and forwards data packets each of which contains a destination address and an originating network-from a LAN or WAN to another. The rauters are layer 3, bridges layer 2, and locating the best route for all data received from another Rauter or last station LAN.

Convert packets from the LAN, packets able to be sent via wide area networks. During sending process, Rauter look and examines the packet destination address and at its own address table, which maintains updated exchanging addresses with other rauters to establish binding paths through the networks that interconnect them. This exchange of information between rauters is done using management protocols owners.

Classification of Routers:

Static: The update tables is manually.

Dynamic: The update of the tables is automatically. There is dynamic Interior Protocols (IGP) for example RIP (Routing Information Protocol that communicates different systems belonging to the same logical network. They have tables and dynamic routing information is exchanged as needed. The tables contain where to go to different destinations and the number of hops that have to be made. This technique allows a maximum of 14 hops so this one is not scalable protocol. Another example is OSPF (Open Shortest Path First Routing). It is designed to minimize traffic routing, allowing full authentication of messages sent. Rauter each have a copy of the network topology and all copies are identical. Each Rauter distributes information to its adjacent Rauter. Each team builds a routing tree regardless, this protocol are scalable. There is Dynamic External Protocols (EGP): Exterior Gateway Protocol (EGP) This protocol allows routers to connect two independent systems that exchange update messages. This protocol is used to establish a source-destination paths.

There is tree Routed protocols:
IPX   IP   AppleTalk

IPV-6, have 128 bits. Incorporates encryption standard but security remains a problem. RFC support IP version 4 for easy migration.

Internet Protocol (IP) is a connectionless protocol. Version 4 is 32 bits. Have two limitations because this thinking developed as an open system and for a specific use. The first is that it is not safe and the second is not scalable.

To understand the operation of IP version 4 will analyze the class hierarchy. This was designed in the likeness of the e-mail; in a letter there are two elements to define the address: First and second street number of the house.

IP version 4 has two components, the network part in the left side and the host part to the right. It is important to remember these items for future reference in this regard will number the steps.

Step 1: The class A is 8 bit networks and 24 for host. Class B has 16 bits for networks 16 to hosts. Class C have 24 bits networks and 8 for host .

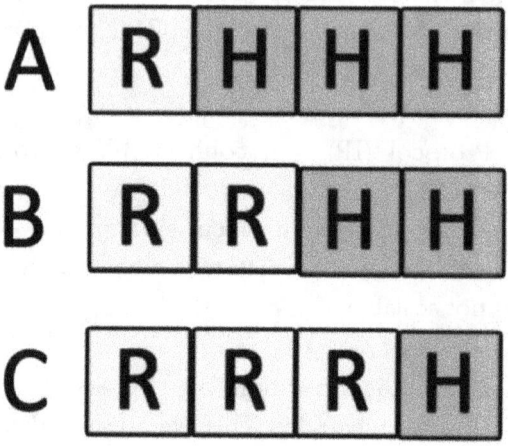

Figure 25: Class Hierarchy

Step 2: The IP address is represented in the decimal system. If we represented the blocks as follows:

$$X.X.X.X$$

An IP address would be represented by X in any number of the range 0-255:

$$2^8 = 256 \quad \leq X \geq 255$$

2 Because the binary system is 0 and 1. 8 Because numbers of bit result 256 combinations.

Example of an IP address: 10.200.30.64.

Step 3. The X on the left is the most significant value and is the defining class. For class A need only one bit, for the B two and C three, therefore: :

A = 0   (-1)

B = 10   (-2)

C =110   (-3)

Step 4: They take only the first 4-bit, 16 combinations that would allow us to see the behavior of all classes taking into account that in the beginning we completed at 0 and end at 1 for each class.

Summary table below:

A= 1-126   B=128-191   C=192-223.

Step 5. Total Network and host for each class:

Class Ar = $2^{8-1=7}$ => X-2 = 126   (Step 3)

Class Ah= 2^{24} => X-2 = 16,777,214 (16M)

Class Br= $2^{16-2=14}$ => X-2 =16,382 (Step 3) 16K

Class Bh= 2^{16} => X-2 =65,534 (65K)

Class Cr= $2^{24-3=21}$ => X-2 = 2,097,150 (Step 3) 2M

Class Ch= 2^8 => X-2 = 254

Clase	128	64	32	16	8	4	2	1	Valor Decimal	Tabla convierte de bin a decimal
	0	0	0	0	0	0	0	0	0	Reservado, base
	0	0	0	1						Todas las clases.
	0	0	1	0						
A	0	0	1	1						Clase A=1-126
	0	1	0	0						
	0	1	0	1						
	0	1	1	0						
	0	1	1	1	1	1	1	1	127	Reservado, Broadcast
	1	0	0	0	0	0	0	0	128	
B	1	0	0	1						
	1	0	1	0						Clase B=128-191
	1	0	1	1	1	1	1	1	191	
C	1	1	0	0	0	0	0	0	192	
	1	1	0	1	1	1	1	1	223	Clase C=192-223
D	1	1	1	0						
E	1	1	1	1						
D Multicast	1	1	1	0	0	0	0	0	224	
	1	1	1	0	1	1	1	1	239	
Experimentos	1	1	1	1	0	0	0	0	240	
	1	1	1	1	1	1	1	1	255	

Figure 26. Class Hierarchy

In this universe there are private network addresses for private use can not be published on the Internet:

Fron 10.0.0.0 to 10.255.255.255

Fron 172.16.0.0 to 172.31.255.255

Fron 192.168.0.0 to 192.168.255.255

Subnet mask (subnetting). The subnetting is used to provide scalability to IP version 4. It is a method that separates the host from networks. It is represented in decimal example: 255.0.0.0 and binary 8 having both the same meaning connotations 8 bit networks and 24 for host.

Case 1: Default Mascara. There is no difference between the concept of class hierarchy and the resulting value that defines the mask.

10.1.2.3 / 8 = 10.1.2.3 255.0.0.0.

(Class A 8 bit Network, 24 bit Host) Network=126 Host=16 Mill.

172.16.2.3 / 16 = 172.16.2.3 255.255.0.0.

(Class B 16 bit Network, 16 bit Host)

192.168.2.3 / 24= 192.168.2.3 255.255.255.0

(Class C 24 bit Network, 8 bit Host)

Case 2: Super Network. The mask affects the original IP Host increasing and decreasing Networks.

10.1.2.3 / 7 = 10.1.2.3 254.0.0.0. (Class A 7 bit Network, 25 bit Host) Networks=62 {[(126+2)/2]-

2} Host=32 M. As it is a bit difference decreases and increases half double.

Case 3: Sub Networks. The mask affects the original IP Networks increasing and decreasing Host.

10.1.2.3 / 9 = 10.1.2.3 255.128.0.0. +(Class A 9 bit Network, 23 bit Host) Networkes=254 {[(126+2)x2]-2} H=8 M

As a bit of difference increases and decreases double or half. The sub Networks Class is the most used in the industry so we will analyze all possible.

C = 192.168.1.2 / 24.

From the above example C can only get 6 host bit for subnetting, should be Class B = 14 and A = 22.

We will analyze the sub network Issue 3, see table below. We calculate the jumps there are two ways for this:

First the number of hops is determined:

256-224 (mask in decimal) = 32 is the number of hops.

#	Mask	Prefix	Sub-Network	Hosts
1	255.255.255.0	24	1	2^8=256-2=254
	255.255.255.128	25	0	2^7=128-2=126
2	255.255.255.192	26	2	2^6=64-2=62
3	255.255.255.224	27	6	2^5=32-2=30
4	255.255.255.240	28	14	2^4=16-2=14
5	255.255.255.248	29	30	2^3=8-2=6
6	255.255.255.252	30	62	2^2=4-2=2
	255.255.255.254	31	126	2^1=2-2=0

Second: 128 64 32 16 8 4   2 1 = Conversion Table

1   1 1 1 0 0   0 0 = 224

The last bit is turned on hop count. The Networks are: 0, 32, 64, 96, 128, 160, 192 y 224 (8-2=6)

Is the first usable Network 32.

Base=192.168.1.32/27

Broadcast   =192.168.1.63/27

It's the ultimate jump or next -1 (64-1)

Range 1   192.168.1.33/27 Base+1 (32+1)

Rango 2   192.168.1.62/27 Broadcast-1 (63-1)

La última Network usable es 192.

Base   192.168.1.192/27

Broadcast   192.168.1.223/27

Range 1   192.168.1.193/27

Range 2   192.168.1.222/27

Exercise: Extract all sub Networks of six possible cases, remember that the first method is jumps to obtain Broadcast Base and Range. When you master this method is not yet prepared to successfully analyze an IP al Network or are 6 possible techniques but do not need to know them all. Knowing the jumps discussed above and combining with AND method that will be explained later, we have the ideal place to discuss any IP successfully combined without hesitation. Before exercise should combine the two mechanisms separately. AND Method. The mask multiplies the IP and resulting value is the desired super Network or sub network. The AND Logic is 1 only when both BIT are 1 other multiplications are zero

$$1x1=1, 0x0=0, 1x0=0, 0x1=0.$$

{1} 10.33.62.4 / 10

{2} X1.X2.X3.X4.

{3} 11111111.11000000.0…=10 =255.192.0.0

{4}   128 64 32 16 8 4 2 1 (X2)

{5}   0   0   1   0   0   0   0   1 X2= 33

{6}   1   1   0 0 0 0 0   192

{7} B 0   0   0 0 0   0   0 0 .0.0 =10.0.0.0

{8} BT 0001111. 11111111.11111111=10.63.255.255

{9}R1 00000000. 00000000.00000001=10.0.0.1

{10} R2 0001111. 11111111.11111110=10.63.255.

Between 64 and 32 must place a slash, where the mask separates Networks the host. On the left side of the line the values are equal after the multiplication. The right to apply the following rule:

Base: All bit after the line to the right should be 0.

Broadcast: Every bit after the line to the right should be one.

First IP Range: Base + 1 = R1

Last IP range: Broadcast - 1 = R2

Exercise: Taking the example of the AND as a standard by the 10 steps in the same order each case will be different but the same IP with the following masks / 11, 5, 14, 18.

Complement Mask (Wild Card Mask). This is generally used to create access lists and dynamic configuration protocols.

> 255.255.255.255   Univers.

> - 255.255.255.248   Sub NetMask.

> _____

>   0 . 0. 0. 7   Wild Card Mask.

According to prior learning, this Wild Card Mask affects eight Networks 0-7.The Wild Card Mask IP affects the result of the sum OR logic. Internet security aims to mitigate unauthorized access. Basically you need to configure access lists once.

Security policies are defined. Managing the logical OR has a higher level of complexity at which we will discuss 7 cases as examples, which allow us to obtain a satisfactory and accurate results throughout the universe in any access list (ACL).

Case 1: When the Wild card mask = 1 for odd IP the result is the number -1. An pair IP or 0 the result will be the number + 1.

Acces-list 80 permit IP   Wild Card Mask= Answered

Acces-list 80 permit 10.1.1.1   0.0.0.1= 10.1.1.0 y 1

Acces-list 80 permit 10.1.1.2   0.0.0.1= 10.1.1.2 y 3

Acces-list 80 permit 10.1.1.9   0.0.0.1= 10.1.1.8 y 9

Acces-list 80 permit 10.1.1.8   0.0.0.1= 10.1.1.8 y 9

Case 2: When the Wild Card Mask is continue in order from right to left except the especial case number one.

3=0000001,   7=00000111,   15=00001111,
31=00011111,   63=00111111,   127=01111111,
255=11111111.

Acces-list 80 permit 10.1.1.1   0.0.0.7= 10.1.1.0-7
Acces-list 80 permit 10.1.1.20 0.0.0.7= 10.1.1.16-23
Acces-list 80 permit 10.1.1.20 0.0.0.15= 10.1.1.16-31
Acces-list 80 permit 10.1.1.20 0.0.0.255= 10.1.1.any
(0-255)   Acces-list   80   permit   0.0.0.0

255.255.255.255= any

Case 3: When the Wild card mask = 0 or any couple, except 254 the result is the original IP.

Acces-list 80 permit 10.1.1.1   0.0.2.2= 10.1.1.1

Acces-list 80 permit   255.255.255.255   0.0.0.0 =255.255.255.255

Caso 4: When the Wild card mask=254.

Acces-list 80 permit 10.1.1.1   0.0.2.254= 10.1.1. odd

Acces-list 80 permit 10.1.1..0   0.0.0.254 =10.1.1.pars

Case 5: When account is taken the order in which the instructions are listed, in this case does the former.

Acces-list 80 permit 10.1.1.1 0.0.0.0= 10.1.1.1 Execute this line

Acces-list 80 Deny   10.1.1.1 0.0.0.0= 10.1.1.1 Ignore this line

Caso 6: There is an implicit Deny end of each list. Although the intention is to deny ip 10.1.1.1 is denying all.

Acces-list 80 Deny   10.1.1.1 0.0.0.0= 10.1.1.1

Acces-list 80 Deny ALL (This line is invisible in all

ACLs).

Case 7: Where the Wild Card is any number that is not mentioned above is because it has particular importance for a list that defines interesting level of security.

> Ping. Echo Request and Echo Replay.
> Measures the connectivity and latency.
> Line quality.
> Tracet: Displays status of links.

Ping. It is a test that determines whether the source could reach the destination. Create and manage a message from time expires if the lifetime of the message. It also determines if the IP header is wrong.

To measure the speed of response of different servers of the country or world. It is very useful to detect if there is a bottleneck. Measures the total time from our PC to a server. Req Echo / Echo Reply proves that the destination is active. Testing the Network is working properly. The most important thing we can know if there is latency in a Network, an acceptable answer is between 0 and 60 milliseconds.

Tracert: It is a more professional way of taking

measurements. You can do this with a similar mechanics to that noted with Ping, Ping just replacing the word by word Tracert. The Tracert detailed measurement times for each leg of the various links to get to the selected server.

6. TRASPORT LEVEL

Transport Layer: This layer is responsible for the integrity of the data. Your task is to make the data transport is done safely and economically, between the destination and origin, it does not depending on the amount of physical Networks that are in use. To achieve this, the transport layer uses all services provided by the Network Layer. It has two basic protocols: Transmission Control Protocol, TCP (Transmission Control Protocol) and Protocol program data using UDP (Use large protocol data). TCP is a connection-oriented protocol that is used ACK and UDP connectionless.

To define the data flow (data flow) of the transport layer is necessary to know the following terms: Window (Windowing): The amount of data that can be sent before receiving an acknowledgment (ACK = Acknolagement). Sets the size of the segment.

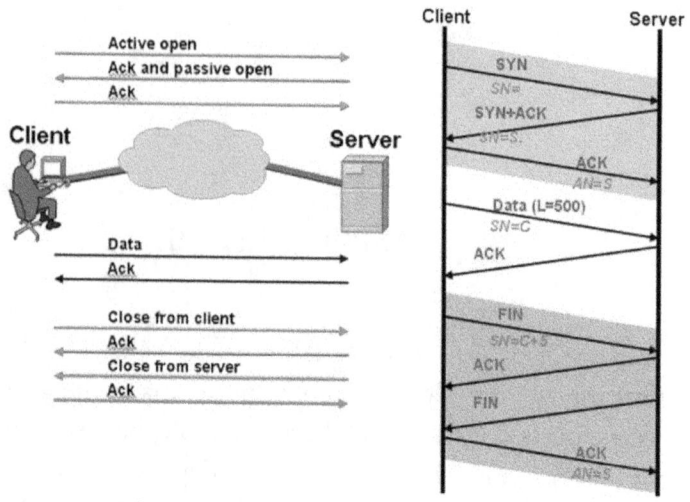

Figure 27: Acknolagement (ACK)

If the window is too small the data would perform poorly and have very large if many errors. Ideally, a window of size 3 is the default value.

Capacity of equipment (Buffering): This concept relies on fundamental benefits for using the central processing unit (CPU) that performs arithmetic operations. The programs perform automated tasks such as operating systems have three basic states: ready, blocked and running, this means that we can interrupt the CPU at a given time. With this feature we can further exploit the CPU through the performance.

Classification of services:

Temporary disk storage (Spooling) temporary memory storage (buffering). Magnetic tape storage (Line Off).

Off The Line is the most used at the CPU, companies often use it to backup of daily transactions. The backup tapes are done at night since the tape continuously interrupted CPU. During the day, if it joins the daily work, the processing would be very slow, unless a separate server is used for these purposes.

Congestion can be defined as the bottleneck when Networks cannot handle the traffic coursing through them. Latency plays a very important role in the environment and is more significant in the Wan Networks. The data flow in the transport level depends on the relation between the windowing, bufferinf and congestion. Analyze the following cases: The B terminal equipment are very good Buffering, the b small equipment.

Case1: Good Buffering, Good bandwidth. This

accept large size Windowing, it would be ideal.

B   B

Figure 28. Good Bandwidth

Case 2: Good Buffering,, low bandwidth Windowing not bear large size, this company should invest more in bandwidth.

B   B

Figure 29. Congestion

Case 3: Low Buffering,, low-bandwidth windowing bear this small, would be ideal.

b   b

Figure 30. Acceptable bandwidth

Case 4: Equipment slow terminals, this much bandwidth Windowing bear small, this company should Network expenses bandwidth.

b   b

31. Waste of bandwidth

Case 5: different terminal equipment, good bandwidth Windowing bear this small, we must take care of the right side. If the small b gets larger windowing, Rx would overrun problems (inability to receive buffer). If the small b like to convey large windowing, Tx underrun have problems (inability to transmit buffer).

B  b

Figure 32. Over Run/Under run

7. INTEROPERABILITY

Interoperability. Levels 5, 6 and 7. Comparing TCP / IP with the OSI model. TCP / IP is a protocol that encompasses a family of communication protocols (over 100), which determine the rules for sending and receiving data over the Networks. This comparison will give us an overview of the operation of the protocol used over TCP / IP world. The level 1 and 2 of the OSI layer is one of TCP / IP and levels 5, 6 and 7 are 4 TCP / IP.

As shown in the following Figure follows TCP / IP protocols oriented handles the connection and not allowing it to be a very versatile protocol.

Next we define some protocols as OSI Layer 7 and 4 According to TCP / IP protocols are called both applied:

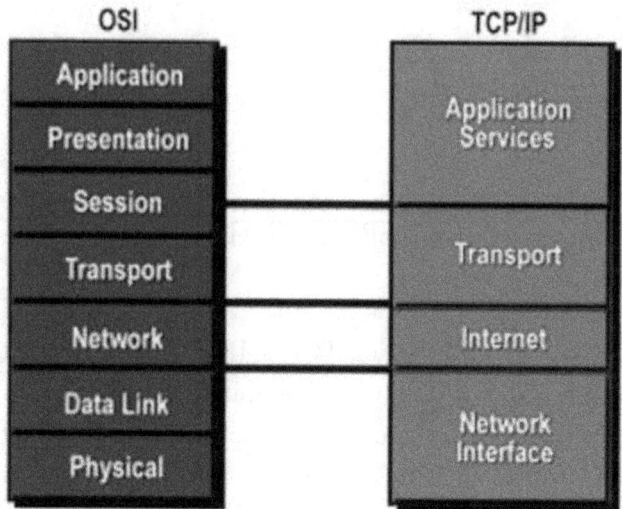

Figure 33: The Keys Similarities between the Network layer and Transport

HTTP, the Hypertext Transfer Protocol (Hypertext Transfer Protocol). Is the set of rules, or protocols, which govern the Hypertext Transfer between two or more computers. The World Wide Web encompasses the universe of information that is available via HTTP.

Hypertext is a special way of coding using a standardized system called Hypertext Markup Language, HTML (Hypertext Markup Language). HTML code is used to create links (links). These links may be textual or Figures and are made by a simple touch of the button "mouse" with the simple touch all other HTML resource type, such as documents, Figures, text files, animation, and sound are available.

Telnet. Remote Access. It is a protocol or set of rules that allows a computer to connect to another. This process is also known as a remote login:

1. Connect to a computer "remote"
2. Protocols
3. Terminal Emulation
4. Based on Client / Server
5. Access to resources on other machines

The user's computer, which initiates the connection, it is known as a local computer, and the computer you are connecting, which accepts the connection is remote or host computer. The remote computer can be physically accommodated in the next room, in another city, or another country. Once connected, the user's computer emulates the remote computer. When the user types commands at the keyboard, these are executed on the remote computer. The monitor shows the user what is happening on the remote computer during the telnet session. The procedure to connect to a remote computer is configured depends on how your Internet access. Once the connection to the remote computer has been established, instructions or menus may appear. Some remote machines require the user to have an account on that machine, and questioned by a username and password.

Any resources such as library catalogs, are available via telnet without username and password.

Telnet operates on the principle of client / server. The

local computer using a telnet client program to connect and display data on the screen of the local computer. The remote host computer or server uses a telnet program to accept the connection and send responses to requests for information back to the local computer.

Telnet allows users to access Internet resources on other computers Network the world. A variety of resources are available through telnet. In short, Telnet is the protocol that allows a computer to make a connection to another computer.

Although some computers may require an account name and password, most computers allow users to access resources stored on them without an account name and password. Telnet provides access to many resources of the world as al library catalogs, databases, and other Internet tools and connection oriented FTP, File Transfer Protocol, FTP applications. It is the technology that lets you connect with your via the internet to update and maintain your website, for example. Use ACK.

TFTP Trivial File Transfer Protocol, TFTP this is a non-connection-oriented protocol. One of the possibilities of the Internet is to copy and save files from one computer to another via modem.

SMTP, Simple Protocol for Post sent, SMTP. It is the protocol used to send mail. The same also holds for the server to receive and refer to the appropriate box.

DNS, Domain Name Server, DNS is a text name added to the server name to form a unique Internet hostname. Every computer connected to the Internet has (at least) one associated identification number: your IP number. In any Internet number is not repeated in different computers.

Each computer has usually a name. Groups of computers are organized into domains allowing different computer with the same name as long as they are in different domains.

It is the responsibility of management to provide this service domain name resolution for the connected computers. This involves two tasks: installing name servers and updating records in the managed domain names.

Good management can make DNS more reliable and efficient operation of the domain. Often this also involves configuration tasks computers domain clients.

SNMP, Simple Management Network Protocol (Simple Network Management Protocol). Designed in the 80s, its main objective was to integrate the management of different types of Networks with a simple design that caused little overhead on the Network. SNMP operates in the application layer, using the UDP transport protocol (not connection oriented), so ignore the underlying hardware on which it runs. The management is done at the IP level, so you can control devices that are connected to any Network accessible from the Internet, not just those in the local Network itself. Obviously, if any of the routing devices with the remote control device is not working properly, it will not be possible to monitor or reconfigure.

The SNMP protocol is composed of two elements: the agent (agent) and the manager (manager). It is a client-server architecture in which the agent is the server and the manager is the client.

The agent is a program that is running in each node Network to be managed or monitored. Provides an interface to all the elements that can be configured. These items are stored in a data structure called managed information base, MIB (Management Information Base). Represents the server, to the extent that the information is managed and waits for commands from the client.